Focus On

Speed, Ratio and Proportion

KUMON

1 Car A traveled 70 miles in 2 hours, and car B traveled 120 miles in 3 hours. Answer the questions below. (12 points per question)

① How many miles did car A travel in one hour?

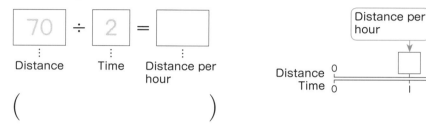

70	÷	2	=	
Distance		Time		Distance per hour

(   )

Distance per hour | Distance covered in two hours

Distance 0 _____ 70 (mi.)
Time 0 ___ 1 ___ 2 (h)

② What is car A's speed? Answer in mph.

Miles per hour (mph) is the distance in miles covered in one hour. Therefore, the answer to question ① is equal to the answer to question ②.

(mph)

③ How many miles did car B travel in one hour?

120	÷	3	=	

(   )

Distance per hour

Distance 0 _____ 120 (mi.)
Time 0 ___ 1 ___ ___ 3 (h)

④ What is car B's speed? Answer in mph.

(mph)

⑤ Which car is faster, A or B?

(   )

2 Answer each question below in mph.

① If a train travels 240 miles in 4 hours, what is its speed?

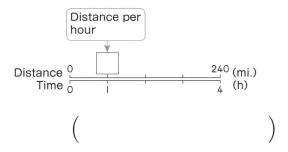

(   )

② If a bird flies 80 miles in 2 hours, what is its speed?

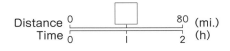

(   )

③ If an express train travels 600 miles in 5 hours, what is its speed?

(   )

④ If an airplane flies 2,440 miles in 4 hours, what is its speed?

(   )

2 Speed II

1 You biked 1,500 meters in 6 minutes. Answer the questions below.

15 points per question

① How many meters did you bike in one minute?

1500 ÷ 6 = ☐

Distance   Time   Distance per minute

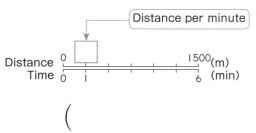

(   )

② What was your speed while on the bike? Answer in meters per minute.

> Meters per minute (m/min) is the distance in meters covered in one minute. Therefore, the answer to question ① is equal to the answer to question ②.

(m/min)

2 Answer each question below in m/min.

10 points per question

① If a man walks 600 meters in 8 minutes, what is his speed?

(   )

② If a motorcycle travels 9,000 meters in 12 minutes, what is its speed?

(   )

3 A pigeon flies 210 meters in 7 seconds. Answer the questions below.

① How many meters does the pigeon fly in one second?

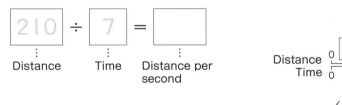

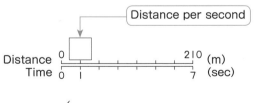

(   )

② What is the speed of the pigeon? Answer in m/sec.

> Meters per second (m/sec) is the distance in meters covered in one second.
> Therefore, the answer to question ① is equal to the answer to question ②.

(m/sec)

4 Answer each question below in m/sec.   10 points per question

① If a race car travels 930 meters in 15 seconds, what is its speed?

(   )

② If a tuna swims 360 meters in 24 seconds, what is its speed?

(   )

3 Speed Ⅲ

1 An airplane flies 3,024 kilometers in 3 hours. Answer the questions below.

(20 points per question)

① What is the airplane's speed in kilometers per hour?

$$\boxed{3024} \div \boxed{3} = \boxed{}$$

⋮ Distance   ⋮ Time

Distance ÷ Time = Speed

Distance $\begin{array}{c} 0 \quad \boxed{} \quad 3024 \text{ (km)} \\ \hline \end{array}$
Time $\quad 0 \quad\quad 1 \quad\quad\quad 3$ (h)

(km/h)

② What is the airplane's speed in kilometers per minute?

$$\boxed{} \div \boxed{60} = \boxed{}$$

1 hour = 60 minutes. Therefore if you divide the kilometers per hour by 60, you get the kilometers per minute.

(km/min)

③ What is the airplane's speed in meters per second?

$$\boxed{} \text{ km} = \boxed{} \text{ m}$$

$$\boxed{} \div \boxed{60} = \boxed{}$$

1 kilometer equals 1,000 meters, You must convert the kilometers into meters to calculate the meters per second.

1 minute = 60 seconds. Therefore if you divide the meters per minute by 60, you get the meters per second.

(   )

2 A race car travels 1,260 kilometers in 5 hours. Answer the questions below.

① What is the race car's speed in kilometers per hour?

( )

② What is the race car's speed in kilometers per minute?

( )

3 An express train travels 270 kilometers per hour. Answer the questions below.

① What is the express train's speed in kilometers per minute?

( )

② What is the express train's speed in meters per second?

( )

4

Speed IV

date / / score /100

1 A man runs 100 meters in 16 seconds. Answer the questions below.

20 points per question

① What is the man's speed in meters per second?

$$\boxed{100} \div \boxed{16} = \boxed{}$$

Distance   Time

Distance ÷ Time = Speed

(   )

② What is the man's speed in meters per minute?

$$\boxed{} \times \boxed{60} = \boxed{}$$

1 minute equals 60 seconds. Therefore if you multiply the meters per second by 60, you get the meters per minute.

(   )

③ What is the man's speed in kilometers per hour?

$$\boxed{} \times \boxed{60} = \boxed{}$$

1 hour equals 60 minutes. Therefore if you multiply the meters per minute by 60, you get the meters per hour.

Don't forget to convert the unit from meters to kilometers.

$$\boxed{} \text{ m} = \boxed{} \text{ km}$$

(   )

8 ©Kumon Publishing Co., Ltd.

2 A dolphin swims 90 meters in 50 seconds. Answer the questions below.

10 points per question

① What is the dolphin's speed in meters per seconds?

( )

② What is the dolphin's speed in meters per minute?

( )

3 The speed of sound is 340 meters per second. Answer the questions below.

10 points per question

① What is the speed of sound in kilometers per minute?

( )

② What is the speed of sound in kilometers per hour?

( )

1 A bus travels 5 I kilometers per hour, and a taxi travels 900 meters per minute. Compare their speeds.

[20 points per question]

① Compare their speeds by converting the bus' speed into meters per minute.

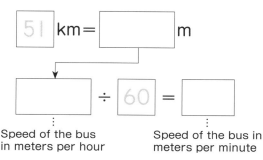

Speed of the bus in meters per hour   Speed of the bus in meters per minute

Convert the unit from kilometers to meters.

Meters per minute = meters per hour ÷ 60

Which is faster, the bus or the taxi?

(   )

② Compare their speeds by converting the taxi's speed into kilometers per hour.

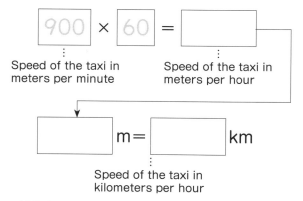

Speed of the taxi in meters per minute   Speed of the taxi in meters per hour

Meters per hour = meters per minute × 60

Speed of the taxi in kilometers per hour

Convert the unit from meters to kilometers.

Which is faster, the bus or the taxi?

To compare different speeds, first convert their units of measurements so they are the same.

(   )

2 A car travels 12.5 meters per second, and a train travels 840 meters per minute. Which is faster? Compare their speeds in the questions below. (20 points per question)

① Compare their speeds by converting the car's speed into meters per minute.

> Meters per minute = meters per second × 60

(   )

② Compare their speeds by converting the train's speed into meters per second.

> Meters per second = meters per minute ÷ 60

(   )

3 A bird flies 50 meters per second, and a pitcher throws a ball 144 kilometers per hour. Which is faster? Compare their speeds by converting both speeds into meters per minute. (20 points per question)

> Meters per minute = meters per second × 60
> Meters per minute = meters per hour ÷ 60

(   )

1 A car is traveling 48 kilometers per hour. Answer the questions below.

(15 points per question)

① How many kilometers will the car travel in one hour?

Kilometers per hour is the distance in kilometers covered in one hour.

(   )

② How many kilometers will the car travel in 2 hours?

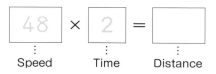

48	×	2	=	
Speed		Time		Distance

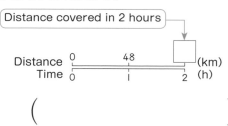

Distance covered in 2 hours

Distance   0   48   ☐ (km)
Time   0   1   2 (h)

(   )

③ How many kilometers will the car travel in 3 hours?

Distance   0   48   ☐ (km)
Time   0   1   3 (h)

(   )

④ How many kilometers will the car travel in 4.5 hours?

(   )

Even when time is shown in decimal form, you can calculate the distance the same way.

2 Answer each question below.

① A storm is traveling 25 kilometers per hour. How far will it go in 3 hours?

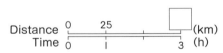

(   )

② A truck is traveling 55 kilometers per hour. How far will it go in 4 hours?

(   )

③ A small plane is traveling 125 miles per hour. How far will it go in 5 hours?

(   )

④ A jet plane is traveling 560 miles per hour. How far will it go in 2.5 hours?

(   )

7 Distance II

1 A man is walking 62 meters per minute. Answer the questions below.

15 points per question

① How many meters will he walk in one minute?

Meters per minute is the distance in meters covered in one minute.

(   )

② How many meters will he walk in 5 minutes?

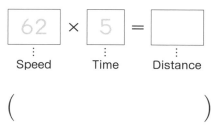

$$ \boxed{62} \times \boxed{5} = \boxed{} $$

Speed   Time   Distance

(   )

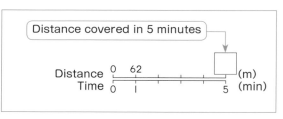

Distance covered in 5 minutes

Distance   0   62   (m)
Time   0   1   5   (min)

2 Answer each question below in meters.

10 points per question

① A cheetah is running 95 meters per minute. How far will it go in 8 minutes?

Distance = Speed × Time

(   )

② A boat is traveling 360 meters per minute. How far will it go in 18 minutes?

(   )

3 A kangaroo is jumping 20 meters per second. Answer the questions below. 〔15 points per question〕

① How far will it go in one second?

Meters per second is the distance in meters covered in one second.

( )

② How far will it go in 7 seconds?

20	×	7	=	
Speed		Time		Distance

( )

4 Answer each question below in meters. 〔10 points per question〕

① A dolphin is swimming 14 meters per second. How far will it go in 9 seconds?

( )

② A jet plane is flying 320 meters per second. How far will it go in 24 seconds?

( )

1 A storm is traveling 35 kilometers per hour. Calculate the time that it takes to travel 280 kilometers by answering the questions below.

[15 points per question]

① Divide the distance (280 kilometers) by the speed (35 kilometers per hour).

280 ÷ ☐ = ☐   (   )

② Fill in the missing measurements below, and answer the number sentence to find out the time it takes the storm to travel 280 kilometers.

☐ ÷ ☐ = ☐
Distance   Speed   Time

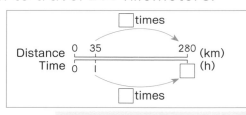

Distance ÷ Speed = Time

(   )

③ How long will it take to travel 490 kilometers?

☐ ÷ ☐ = ☐
Distance   Speed   Time

(   )

④ How long will it take to travel 630 kilometers?

(   )

2 Answer each question below in hours. (10 points per question)

① A monorail is traveling 40 kilometers per hour. How long will it take to travel 200 kilometers?

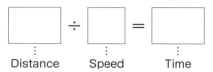

(   )

② A bus is traveling 45 kilometers per hour. How long will it take to travel 540 kilometers?

(   )

③ A bicycle is traveling 15.5 kilometers per hour. How long will it take to travel 62 kilometers?

(   )

④ A boat is traveling 24 kilometers per hour. How long will it take to travel 156 kilometers?

(   )

9 Time II

date ／ ／ score ／100

1 Answer each question below.

15 points per question

① A car is traveling 9 kilometers per minute. How long will it take to travel 234 kilometers?

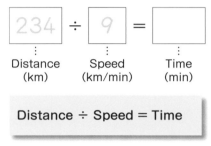

Distance ÷ Speed = Time

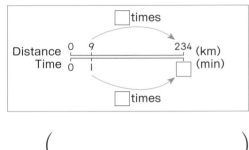

( )

② A man is running 55 meters per minute. How long will he take to run 2,530 meters?

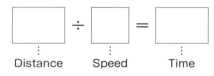

( )

③ A bus is traveling 500 meters per minute. How long will it take to travel 8,500 meters?

( )

④ A rocket is traveling 450 kilometers per minute. How long will it take to travel 7,200 kilometers?

( )

18 ©Kumon Publishing Co., Ltd.

2 Answer each question below.

10 points per question

① A giraffe is walking 14 meters per second. How long will it take to walk 770 meters?

	÷		=	
Distance		Speed		Time

Distance ÷ Speed = Time

(   )

② A horse is walking 21 meters per second. How long will it take to walk 441 meters?

(   )

③ A mortorcycle is traveling 12 meters per second. How long will it take to travel 456 meters?

(   )

④ A man is swimming 0.8 meter per second. How long will he take to swim 40 meters?

(   )

10 Comparing Speeds

1 Copy machine A copies 4,800 sheets per hour. Copy machine B copies 450 sheets in 5 minutes. Copy machine C copies 1.4 sheets per second. Answer the questions below.

15 points per question

① How many sheets can copy machine A copy in one minute?

I hour = ☐ minutes

You can compare the speeds by converting all the copy machine speeds to sheets per minute.

$4800 \div \boxed{} = \boxed{}$

( )

② How many sheets can copy machine B copy in one minute?

$450 \div 5 = \boxed{}$

( )

③ How many sheets can copy machine C copy in one minute?

I minute = ☐ seconds

$1.4 \times \boxed{} = \boxed{}$

( )

④ Which copy machine is fastest, A, B or C?

( )

2 Pump A pumps 160 gallons of water in 32 minutes.
Pump B pumps 72 gallons of water in 15 minutes.
Which pump is faster, A or B?

Calculate the speed (amount of water per minute) of each pump, then compare.

(   )

3 Tractor A plows 30,000 yards in 2 hours. Tractor B
plows 14,000 yards in 50 minutes. Which tractor is
faster, A or B?

Calculate the speed (yards per minute or hour) of each tractor, then compare.

(   )

1 Write the missing words below. [10 points per question]

① Speed = ☐ ÷ ☐

② Distance = ☐ × ☐

③ Time = ☐ ÷ ☐

2 John walks 2.5 miles per hour. Answer the questions below. [10 points per question]

① If John walks for 2 hours, how far will he walk?

( )

② How long would it take John to walk 7.5 miles?

( )

3 Bicycle A travels 1,200 meters in 5 minutes.
Bicycle B travels 4.5 meters per second. Answer the
questions below. （10 points per question）

① How far does bicycle A travel in one minute?

(   )

② How far does bicycle B travel in one minute?

> Meters per minute = meters
> per second × 60

(   )

③ Which bicycle is faster, A or B?

(   )

4 A jet plane is flying 990 kilometers per hour. Answer
the questions below. （10 points per question）

① How far does the jet plane fly in one minute?

(   )

② How far does the jet plane fly in one second? Answer in meters.

(   )

1 The table to the right shows the number of shots and points scored by three people playing basketball. Answer each question below in decimal form.

Name	Number of shots	Points scored
Andy	10	6
Betty	12	9
George	8	4

① is 10 points, ② and ③ are 20 points each.

① Calculate the ratio of Andy's points to shots.

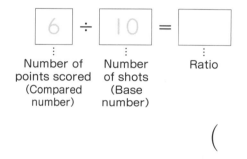

$$\boxed{6} \div \boxed{10} = \boxed{}$$

Number of points scored (Compared number)   Number of shots (Base number)   Ratio

A ratio is the relationship between two numbers.

(   )

② Calculate the ratio of Betty's points to shots.

(   )

③ Calculate the ratio of George's points to shots.

(   )

2 The table to the right shows the membership limit and the number of applicants for the different clubs at Caroline's school. Answer each question below in decimal form.

Club	Membership limit	Applicants
Reading	25	24
Science	30	27
Computer	40	52

(① is 10 points, ② and ③ are 20 points each.)

① Calculate the ratio of the reading club's applicants to its membership limit.

24	÷	25	=	

Applicants (Compared number)　　Membership limit (Base number)　　Ratio

(　　　　　　　　)

② Calculate the ratio of the science club's applicants to its membership limit.

(　　　　　　　　)

③ Calculate the ratio of the computer club's applicants to its membership limit.

(　　　　　　　　)

13 Ratios II

1 Answer the questions below about the rectangle on the right.

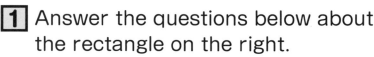

4 cm

7 cm

10 points per question

① The width of the rectangle is 4 centimeters. What is the length?

( )

② Write the ratio of width to length.

You can represent the ratio of width to length by putting a colon (:) between the two numbers.

(4 : 7)

③ Write the ratio of length to width. Answer using a colon.

( )

2 Ribbon A is 9 centimeters, and ribbon B is 20 centimeters. Answer each question below with a colon.

10 points per question

① Write the ratio of ribbon A to ribbon B.

( )

② Write the ratio of ribbon B to ribbon A.

( )

3 Write each ratio below with a colon.

① Write the ratio of 1 liter of water to 2 liters of oil.

(   )

② Write the ratio of 5 pounds of apples to 12 pounds of oranges.

(   )

③ Write the ratio of 17 boys to 14 girls.

(   )

④ Write the ratio of a 21 meter-long backyard to a 50 meter-long field.

(   )

⑤ Write the ratio of 40 yards of wrapping paper to 3 yards of ribbon.

(   )

4 Write each ratio below with a colon. 10 points per question

① The computer club has 8 boys and 7 girls. Write the ratio of boys to girls.

(   )

② You mix 3 tablespoon of oil and 2 tablespoon of lemon juice. Write the ratio of oil to lemon juice.

(   )

1 Write the appropriate number in each box below.

6 points per question

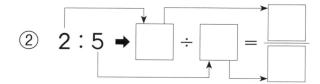

① 1 : 3 ➡ $\boxed{1} \div \boxed{3} = \dfrac{1}{3}$

The value of a ratio is the front number divided by the rear number.

② 2 : 5 ➡ $\boxed{} \div \boxed{} = \dfrac{\boxed{}}{\boxed{}}$

③ 7 : 4 ➡ $\boxed{} \div \boxed{} = \dfrac{\boxed{}}{\boxed{}}$

It is okay if the answer is an improper fraction.

④ 3 : 6 ➡ $\boxed{3} \div \boxed{6} = \dfrac{3}{6} = \dfrac{\boxed{}}{\boxed{}}$

If you can reduce the fraction, reduce it.

⑤ 4 : 2 ➡ $\boxed{} \div \boxed{} = \boxed{}$

2 Write each value of the ratios below.

① 7 : 13

(   )

④ 20 : 5

(   )

② 6 : 15

(   )

⑤ 18 : 8

(   )

③ 9 : 5

The value of the ratio of a:b is $\frac{a}{b}$. If you can reduce the fraction, reduce it.

(   )

3 Pair the ratios that equal each other in the spaces below.

10 points per question

a 1 : 2   b 4 : 5   c 4 : 2

d 8 : 4   e 4 : 8   f 8 : 10

(and   ), (and   ), (and   )

15 Ratios IV

1 Write the appropriate number in each box below.

5 points per question

① 6 : 9 ➡ [12] : [ ] (× 2)

A ratio that is equal to another ratio is called a proportion.

② 6 : 9 ➡ [2] : [ ] (÷ 3)

③ Use the ratios from questions ① and ②,

6 : 9 = [12] : [ ] = [2] : [ ]

The number sentence above is true because the three ratios are equal.

2 Write the appropriate number in each box below.

5 points per question

① 5 : 4 = [ ] : 12 (× 3)

③ 10 : 20 = 1 : [ ] (÷ 10)

② 1 : 3 = [ ] : 12

④ 14 : 4 = 7 : [ ]

3 Reduce each ratio to its simplest form. 5 points per question

① 2 : 4

(   )

③ 12 : 10

(   )

② 6 : 4

(   )

④ 6 : 15

(   )

4 Choose the ratios equal to 4 : 8 from the box.

15 points for completion

a 8 : 10	b 2 : 4	c 20 : 30
d 36 : 72	e 1 : 2	f 6 : 12

(   )

5 Write the appropriate number in each box below.

5 points per question

① 4 : 3 = ☐ : 15

④ 24 : 18 = 8 : ☐

② ☐ : 2 = 3 : 1

⑤ 10 : 25 = ☐ : 5

③ 5 : ☐ = 20 : 16

⑥ ☐ : 9 = 7 : 63

1 Write the appropriate number in each box below.

10 points per question

① $2 : 6 =$ [I] : [ ]

(÷ 2 above, ÷ 2 below)

> The quickest way to reduce a ratio to its simplest form is by dividing both numbers in the ratio by their Greatest Common Factor (GCF) —in this case, it's 6.

② $35 : 21 =$ [ ] : [ ]

(÷ 7 above, ÷ 7 below)

③ $12 : 18 =$ [ ] : [ ]

(÷ 6 above, ÷ 6 below)

④ $0.8 : 1.2 =$ [8] : [ ] $=$ [ ] : [ ]

(× 10 above, × 10 below, ÷ 4 above, ÷ 4 below)

> When the ratio includes decimals, you can convert them into whole numbers first, and then reduce the ratio to its simplest form.

⑤ $\dfrac{1}{5} : \dfrac{2}{3} =$ [ ] : [ ]

(× 15 above, × 15 below)

> When the ratio includes fractions, multiply each number by the Least Common Multiple (LCM) of the denominators in order to make the fractions into whole numbers.

2 Reduce each ratio below to its simplest form.

① 5 : 15

( )

② 100 : 30

( )

③ 24 : 16

( )

④ 200 : 80

( )

⑤ 0.6 : 0.7

( )

⑥ 0.6 : 2.4

( )

⑦ 1.2 : 3.6

( )

⑧ $\frac{2}{5} : \frac{3}{5}$

( )

⑨ $\frac{1}{4} : \frac{1}{3}$

( )

⑩ $\frac{3}{4} : \frac{1}{6}$

( )

date　　／　／　　score　　／100

1 The ratio of a rectangle's length to width is 2 : 3.
Answer the questions below.

10 points per question

① If the length is 8 centimeters, how long is the width?

　(1) Solution I : using the relationship of the equal ratios

　　　If the width is ◎ cm, 2 : 3 = 8 : ◎

$$2 : 3 = 8 : ◎ \qquad \longrightarrow \qquad ◎ = 3 \times \boxed{} = \boxed{}$$

（×4 above, ×☐ below）

　　　　　　　　　　　　　　　　　　　　　　（　　　　　）

　(2) Solution 2 : using the value of the ratio

$$\text{The width} = 8 \times \dfrac{3}{2} = \boxed{}$$

　　　　　　　　　　　　　　　　　　　　　　（　　　　　）

② If the width is 24 centimeters, how long is the length?

　(1) Solution I : using the relationship of the equal ratios

　　　If the width is ◎ cm, 2 : 3 = ◎ : 24

$$2 : 3 = ◎ : 24 \qquad \longrightarrow \qquad ◎ = 2 \times \boxed{} = \boxed{}$$

（× above, ×☐ below）

　　　　　　　　　　　　　　　　　　　　　　（　　　　　）

　(2) Solution 2 : using the value of the ratio

$$24 \times \dfrac{}{3} = \boxed{}$$

　　　　　　　　　　　　　　　　　　　　　　（　　　　　）

2 The ratio of Jacob's older sister's ribbon to his younger sister's ribbon is 5 : 4. Answer the questions below.

① If the older sister's ribbon is 60 centimeters, how long is the younger sister's ribbon?

$5 : 4 = 60 : \square$

( )

② If the younger sister's ribbon is 60 centimeters, how long is the older sister's ribbon?

$5 : 4 = \square : 60$

( )

3 You're making pancakes and the ratio of sugar to flour is 3 : 8. Answer the questions below.

① If you use 120 grams of sugar, how much flour do you need?

( )

② If you use 720 grams of flour, how much sugar do you need?

( )

1 The tables below show the relationship between time and the distance a slug travels. Write the appropriate numbers that you must multiply by in order to get each proportional ratio.

8 points per box

①

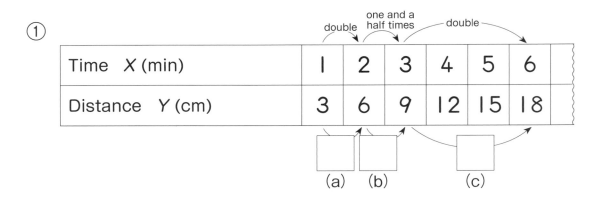

Time　X (min)	1	2	3	4	5	6	
Distance　Y (cm)	3	6	9	12	15	18	

②

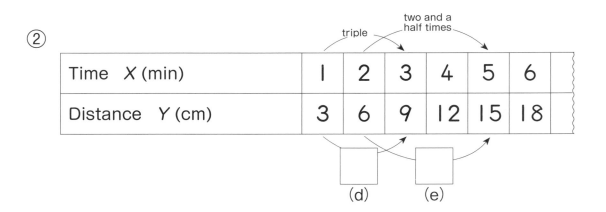

Time　X (min)	1	2	3	4	5	6	
Distance　Y (cm)	3	6	9	12	15	18	

There are two measurements: X and Y. If X is doubled or tripled, Y is doubled or tripled. This is because Y increases in proportion to X.

2 Fill in the following tables with the appropriate numbers. Put a ✓ next to each relationship that is proportional. Put an × if it is not proportional.

① The relationship of the length of one side of a diamond and its perimeter

Length of one side X (cm)	1	2	3	4	5	
Perimeter Y (cm)	4					

② A brother and sister divide 10 sheets of colored paper between each other.

Sheets for the brother X	1	2	3	4	5	
Sheets for the sister Y	9					

③ The relationship between the amount of nails and their weight, which is 2.5 grams each nail

Amount of nails X	1	2	3	4	5	
Weight Y (g)	2.5					

④ The relationship between the ages of a father and child that are both born on the 6th of April

Child's age	1	2	3	4	5	
Father's age	30					

date / /   score /100

1 The table below shows the relationship between the length of steel wire and its weight. Answer the questions below.

10 points per question

Length X (m)	1	2	3	4	5	
Weight Y (g)	6	12	18	24	30	
$Y \div X$						

① Is the relationship between the length and weight proportional or not proportional?

(   )

② Write the appropriate numbers in the bottom row of the table above.

③ Write the appropriate number in the number sentence below.

$$Y = \boxed{} \times X$$

> If Y is proportional to X, we can write
> Y = constant number × X

2 Write the appropriate numbers in the bottom row of the table below, and then write a number sentence that shows the relationship between X and Y.

25 points

One side of square X (m)	1	2	3	4	
Perimeter of square Y (m)	4	8	12	16	
$Y \div X$					

(Y =   )

3 Fill in the following tables below. Then write the number sentence showing the relationship between X and Y.

15 points per question

① The relationship between time and distance traveled by a car going 50 kilometers per hour

Time X (h)	1	2	3	4	5	
Distance Y (km)	50					
Y ÷ X						

(Y =)

② The relationship between height and area of a parallelogram with a base that is 6 centimeters

Height X (cm)	1	2	3	4	5	
Area Y (cm²)	6					
Y ÷ X						

( )

③ The relationship between volume and weight for oil that weighs 0.8 kilogram per liter

Volume X (L)	1	2	3	4	5	
Weight Y (kg)	0.8					
Y ÷ X						

( )

date / / score /100

1 The table below shows the relationship between the amount of nails X and their weight Y (g). Answer the questions below.

8 points per question

Amount of nails X	1	2	3	4	5	
Weight Y (g)		3	6	9	12	15

① Write a number sentence that shows the relationship between X and Y.

( )

② Use the number sentence above to solve for Y in each question below.

(1) X = 10

( )

(2) X = 15

( )

(3) X = 60

( )

(4) X = 200

( )

When we have a number sentence showing the relationship between X and Y, we can calculate the value of X for each Y, too.

2 The table below shows the relationship between time X (h) and distance Y (km) of a speed boat. Answer the questions below. 〔10 points per question〕

Time X (h)	1	2	3	4	5
Distance Y (km)	40	80	120	160	200

① Write a number sentence that shows the relationship between X and Y.

(   )

② Calculate the value of Y when X is 2.5.

(   )

③ Calculate the value of X when Y is 320.

(   )

> How many hours does the speed boat take to travel 320 kilometers?

3 Write a number sentence that shows the relationship between X and Y. 〔10 points per question〕

① The relationship between the length X (m) and weight of an iron bar that weighs 0.6 kilograms per meter Y (kg)

(   )

② The relationship between the length X and perimeter of an equilateral triangle Y

(   )

③ The relationship between the time X (min) and distance covered by a man walking 65 meters per minute Y (m)

(   )

date　　/　/　　score　　/100

1 The table below shows the relationship between time and the distance a slug travels. Use the table to fill in the graph below.

25 points per question

Time X (min)	0	1	2	3	4	5	6	7	8	9	10
Distance Y (cm)	0	2	4	6	8	10	12	14	16	18	20

① Write dots on the graph showing the distance covered according to the time.

② Connect the dots to complete the graph.

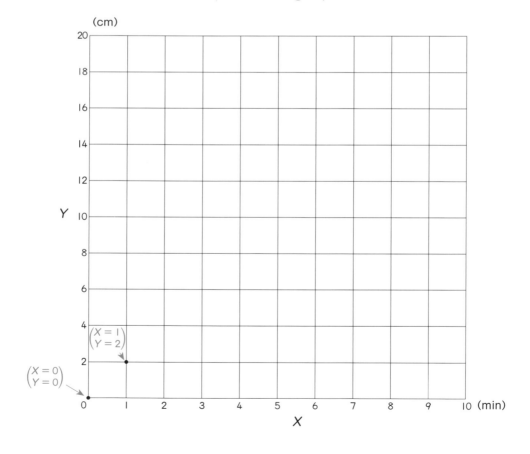

2 A cyclist travels 0.3 kilometer per minute. Answer the questions below.

25 points per question

① Write appropriate numbers in the table below.

Time X (min)	0	1	2	3	4	5	
Distance Y (km)	0	0.3					

② Use the table above, and complete the graph below to show the relationship between time and distance.

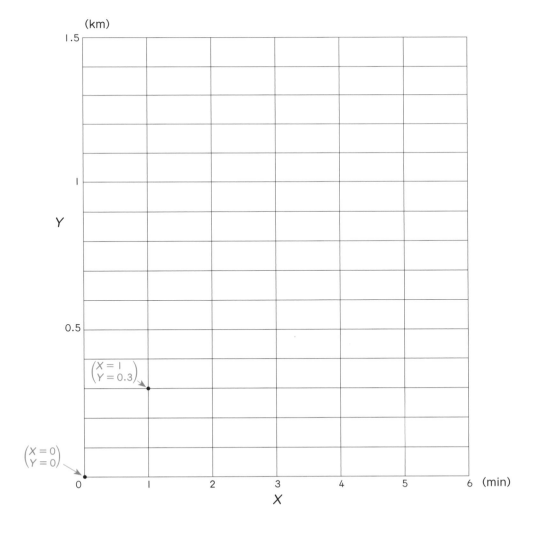

1 The graph to the right shows the relationship between time X (min) and volume of water filling a sink Y (L). Answer the questions below. (10 points per question)

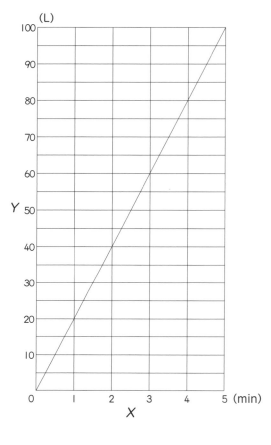

① What is the volume of water in the sink according to each time below?

(1) 2 minutes ( )

(2) 5 minutes ( )

② Write a number sentence that shows the relationship between X and Y.

( )

③ How much time has passed according to each amount of water below?

(1) 20 L (2) 80 L

( ) ( )

④ How much water is in the sink at 2 minutes 30 seconds?

( )

2 The graph to the right shows the relationship between the volume of gas used *X* (gal) and distance traveled *Y* (mi.) by car A and car B. Answer the questions below. 8 points per question

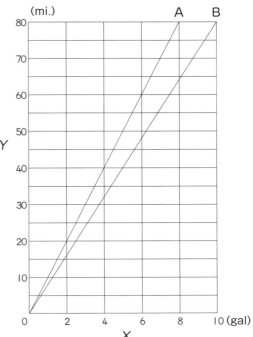

① How far can each car run on one gallon of gas?

(1) Car A

Car A runs 80 miles on 8 gallons of gas.

( )

(2) Car B

Car B runs 80 miles on 10 gallons of gas.

( )

② Write number sentences that show each car's relationship between volume of gas and distance.

A ( )   B ( )

③ You used 25 liters of gas to travel 200 miles. Which car did you drive?

( )

23 Proportion VI

1 You weighed a bag of 25 nails, and it weighed 50 grams. You weighed a second bag of nails, and it weighed 150 grams. Calculate the number of nails. (20 points per question)

① What is proportional to the number of nails?

(   )

② Calculate the number of nails in the second bag in two ways below. Write the appropriate number in each box.

   (1) Solution 1:

      What is the weight of 1 nail? $\boxed{50} \div \boxed{25} = \boxed{}$

      The number of nails in the second bag is $150 \div \boxed{} = \boxed{}$

(   )

   (2) Solution 2:

      How many times larger is the second bag of nails than the first?

      $\boxed{150} \div \boxed{50} = \boxed{}$

      The number of nails in the second bag is also $\boxed{}$ times larger than the first bag.

      The number of nails in the second bag is $25 \times \boxed{} = \boxed{}$

(   )

2 When Jack weighed an iron bar that was 5 centimeters long, its weight was 60 grams. How much would an iron bar that is 24 centimeters long weigh? Calculate the answer using two solutions in questions ① and ②.

20 points per question

① First calculate the weight of one centimeter, and then calculate the weight of 24 centimeters.

Calculate the answer by using the weight of the iron bar in proportion to its length.

(   )

② First calculate how many times longer the 24 centimeters iron bar is, and then calculate its weight.

(   )

1 A 100 cm² square sheet of steel weighs 125 grams. An oval sheet of steel weighs 225 grams. Calculate the area of the oval.

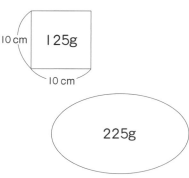

10 cm | 125g | 10 cm

225g

(20 points per question)

① What is proportional to the area?

( )

② Calculate the area of the oval sheet of steel in two ways below. Write the appropriate number in each box.

(1) Solution 1:

What is the area of one gram of the steel sheet?

100 ÷ 125 = [ ]

What is the area of the oval steel sheet?

[ ] × 225 = [ ]   ( )

(2) Solution 2:

How many times heavier is the oval sheet of steel than the square?

225 ÷ 125 = [ ]

The oval sheet of steel is also [ ] times larger than the square sheet.   So the area of the oval sheet is

100 × [ ] = [ ]   ( )

2 Shapes A and B are made with craft paper. Shape A is 24 cm² and weighs 15 grams. Shape B weighs 33 grams. What is the area of shape B? Calculate the answer using two solutions in questions ① and ②. 20 points per question

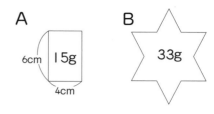

① First calculate the area of one gram of craft paper, and then calculate the area of shape B.

We can calculate the area by using a formula with the proportion.

(   )

② First calculate how many times heavier shape B is than shape A, and then calculate the area of shape B.

(   )

1 Write a number sentence that shows the relationship between X and Y for each sentence below. Then, choose the ratios that are equal to each other.

20 points

a. The relationship between the length of one side X (cm) and the perimeter Y (cm) of an equilateral triangle

b. The relationship between the base X (cm) and height Y (cm) of a 50 cm² parallelogram

c. The relationship between the number of candies X and the weight of one candy Y (g) in a 10 pound bag of candy

d. The relationship between time X (min) and volume of water in the sink Y (L) when 3 liters of water pours from a faucet each minute

(and   )

2 Write a number sentence that shows the relationship between X and Y to each question below. 10 points per question

① The relationship between the length X (m) and the cost Y ($) of ribbon that costs $ 0.70 per meter

(   )

② The relationship between one side X (cm) of a square and its perimeter Y (cm)

(   )

③ The relationship between time X (h) and distance Y (km) of a woman walking 3.5 kilometers per hour

(   )

3 The graph pictured here shows the relationship between the volume of iron and its weight. Answer the questions below.

10 points per question

① Write a number sentence that shows the relationship between X and Y.

( )

② How much does 10 cm³ of iron weigh?

( )

③ How much does 25 cm³ of iron weigh?

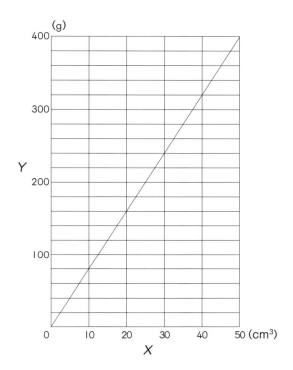

( )

④ What is the volume of 240 grams of iron?

( )

⑤ What is the volume of 600 grams of iron?

( )

1 Lance is training to run the 100 meters race at the track meet. Today he ran 150 meters in 24 seconds. Answer the questions below.

10 points per question

① How many meters per second did he run?

( )

② How many meters per minute could he run?

( )

2 Select the equal ratio from the numbers on the right and write it below.

10 points per question

① 1 : 2 is equal to   2 : 3   2 : 4   2 : 5   2 : 6

( )

② 2 : 3 is equal to   1 : 3   3 : 5   4 : 6   5 : 7

( )

③ 4 : 3 is equal to   6 : 4   8 : 6   10 : 8   12 : 10

( )

④ 5 : 3 is equal to   10 : 5   12 : 4   15 : 6   20 : 12

( )

3 The graph at the bottom of the page shows the relationship between the length of a fabric X (m) and its cost Y ($). Use the graph to answer the questions below.

① Complete this table according to the graph.

Length X (m)	0	1	2	3	4	5	
Cost Y ($)	0	1.40					

② What is Y ÷ X ?

(   )

③ Write a number sentence to show the relationship between Y and X.

(   )

④ How much does 7 meters of fabric cost?

(   )

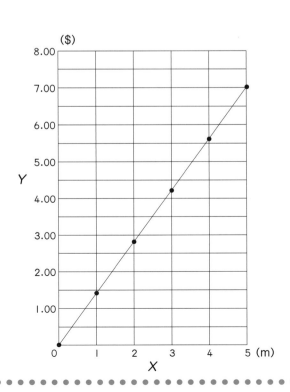

53

date   /   /   score   /100

1 Write the appropriate number in each box so that each pair of ratios are proportional.

5 points per question

① $2 : 3 = 6 : \boxed{}$ ⑥ $3 : 12 = 1 : \boxed{}$

② $4 : 3 = 12 : \boxed{}$ ⑦ $8 : 2 = \boxed{} : 1$

③ $3 : 5 = 15 : \boxed{}$ ⑧ $6 : 8 = 3 : \boxed{}$

④ $3 : 8 = \boxed{} : 24$ ⑨ $25 : 15 = 5 : \boxed{}$

⑤ $5 : 6 = 30 : \boxed{}$ ⑩ $18 : 27 = 2 : \boxed{}$

2 Our plane was flying 480 kilometers per hour and it flew 800 kilometers. How long did our trip take?

20 points per question

(   )

3 The red ribbon at the craft store costs $2.00 per meter. Answer the questions below. 10 points per question

① Complete this table based on the ratio between length and cost.

Length X (m)	0	1	2	3	4	5	
Cost Y ($)	0						

② Complete the graph according to the table above.

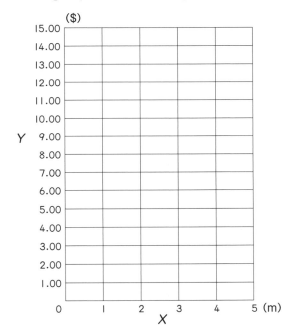

③ Write a number sentence that shows the relationship between length and cost.

( )

1 Speed I
P2-3

1 ① $70 \div 2 = 35$

35 mi.

② 35 mph

③ $120 \div 3 = 40$

40 mi.

④ 40 mph

⑤ Car B

2 ① $240 \div 4 = 60$

60 mph

② $80 \div 2 = 40$

40 mph

③ $600 \div 5 = 120$

120 mph

④ $2440 \div 4 = 610$

610 mph

2 Speed II
P4-5

1 ① $1500 \div 6 = 250$

250 m

② 250 m/min

2 ① $600 \div 8 = 75$

75 m/min

② $9000 \div 12 = 750$

750 m/min

3 ① $210 \div 7 = 30$

30 m

② 30 m/sec

4 ① $930 \div 15 = 62$

62 m/sec

② $360 \div 24 = 15$

15 m/sec

3 Speed III
P6-7

1 ① $3024 \div 3 = 1008$

1,008 km/h

② $1008 \div 60 = 16.8$

16.8 km/min

③ $16.8 \text{ km} = 16800 \text{ m}$

$16800 \div 60 = 280$

280 m/sec

2 ① $1260 \div 5 = 252$

252 km/h

② $252 \div 60 = 4.2$

4.2 km/min

3 ① $270 \div 60 = 4.5$

4.5 km/min

② $4.5 \text{ km} = 4500 \text{ m}$

$4500 \div 60 = 75$

75 m/sec

④ Speed Ⅳ

P8-9

1 ① $100 \div 16 = 6.25$

6.25 m/sec

② $6.25 \times 60 = 375$

375 m/min

③ $375 \times 60 = 22500$

$22500 \text{ m} = 22.5 \text{ km}$

22.5 km/h

2 ① $90 \div 50 = 1.8$

1.8 m/sec

② $1.8 \times 60 = 108$

108 m/min

3 ① $340 \times 60 = 20400$

$20400 \text{ m} = 20.4 \text{ km}$

20.4 km/min

② $20.4 \times 60 = 1224$

1,224 km/h

⑤ Speed Ⅴ

P10-11

1 ① $51 \text{ km} = 51000 \text{ m}$

$51000 \div 60 = 850$

Taxi

② $900 \times 60 = 54000$

$54000 \text{ m} = 54 \text{ km}$

Taxi

2 ① $12.5 \times 60 = 750$

Train

② $840 \div 60 = 14$

Train

3 $50 \times 60 = 3000$

$144000 \div 60 = 2400$

Bird

⑥ Distance Ⅰ

P12-13

1 ① 48 km

② $48 \times 2 = 96$

96 km

③ $48 \times 3 = 144$

144 km

④ $48 \times 4.5 = 216$

216 km

2 ① $25 \times 3 = 75$

75 km

② $55 \times 4 = 220$

220 km

③ $125 \times 5 = 625$

625 mi.

④ $560 \times 2.5 = 1400$

1,400 mi.

⑦ Distance Ⅱ

P14-15

1 ① 62 m

② $62 \times 5 = 310$

310 m

2 ① $95 \times 8 = 760$

760 m

② $360 \times 18 = 6480$

6,480 m

3 ① 20 m

   ② 20 × 7 = 140

   140 m

4 ① 14 × 9 = 126

   126 m

   ② 320 × 24 = 7680

   7,680 m

⑧ Time I
P16-17

1 ① 280 ÷ 35 = 8

   8

   ② 280 ÷ 35 = 8

   8 hours

   ③ 490 ÷ 35 = 14

   14 hours

   ④ 630 ÷ 35 = 18

   18 hours

2 ① 200 ÷ 40 = 5

   5 hours

   ② 540 ÷ 45 = 12

   12 hours

   ③ 62 ÷ 15.5 = 4

   4 hours

   ④ 156 ÷ 24 = 6.5

   6.5 hours

   (6 hours 30 minutes)

⑨ Time II
P18-19

1 ① 234 ÷ 9 = 26

   26 minutes

   ② 2530 ÷ 55 = 46

   46 minutes

   ③ 8500 ÷ 500 = 17

   17 minutes

   ④ 7200 ÷ 450 = 16

   16 minutes

2 ① 770 ÷ 14 = 55

   55 seconds

   ② 441 ÷ 21 = 21

   21 seconds

   ③ 456 ÷ 12 = 38

   38 seconds

   ④ 40 ÷ 0.8 = 50

   50 seconds

⑩ Comparing Speeds
P20-21

1 ① 1 hour = 60 minutes

   4800 ÷ 60 = 80

   80 sheets

   ② 450 ÷ 5 = 90

   90 sheets

   ③ 1 minute = 60 seconds

   1.4 × 60 = 84

   84 sheets

   ④ B

2 A : 160 ÷ 32 = 5

   B : 72 ÷ 15 = 4.8

A

3 A : 2 hours = 120 minutes

30000 ÷ 120 = 250

B : 14000 ÷ 50 = 280

B

⑪ **Review: Speed** P22-23

1 ① Speed = Distance ÷ Time

② Distance = Speed × Time

③ Time = Distance ÷ Speed

2 ① 2.5 × 2 = 5

5 mi.

② 7.5 ÷ 2.5 = 3

3 hours

3 ① 1200 ÷ 5 = 240

240 m

② 4.5 × 60 = 270

270 m

③ B

4 ① 990 ÷ 60 = 16.5

16.5 km

② 16.5 km = 16500 m

16500 ÷ 60 = 275

275 m

⑫ **Ratios I** P24-25

1 ① 6 ÷ 10 = 0.6

0.6

② 9 ÷ 12 = 0.75

0.75

③ 4 ÷ 8 = 0.5

0.5

2 ① 24 ÷ 25 = 0.96

0.96

② 27 ÷ 30 = 0.9

0.9

③ 52 ÷ 40 = 1.3

1.3

⑬ **Ratios II** P26-27

1 ① 7 cm

② 4 : 7

③ 7 : 4

2 ① 9 : 20

② 20 : 9

3 ① 1 : 2

② 5 : 12

③ 17 : 14

④ 21 : 50

⑤ 40 : 3

4 ① 8 : 7

② 3 : 2

14 Ratios Ⅲ

1 ① $1 \div 3 = \dfrac{1}{3}$

② $2 \div 5 = \dfrac{2}{5}$

③ $7 \div 4 = \dfrac{7}{4}$

④ $3 \div 6 = \dfrac{3}{6} = \dfrac{1}{2}$

⑤ $4 \div 2 = 2$

2 ① $\dfrac{7}{13}$ ④ 4

② $\dfrac{2}{5}$ ⑤ $\dfrac{9}{4}$

③ $\dfrac{9}{5}$

3   a and e, b and f, c and d

15 Ratios Ⅳ
P30-31

1 ① $12 : 18$
② $2 : 3$
③ $12 : 18 = 2 : 3$

2 ① 15 ③ 2
② 4 ④ 2

3 ① $1 : 2$ ③ $6 : 5$
② $3 : 2$ ④ $2 : 5$

4   b, d, e, f

5 ① 20 ④ 6
② 6 ⑤ 2
③ 4 ⑥ 1

16 Ratios Ⅴ
P32-33

1 ① $1 : 3$
② $5 : 3$
③ $2 : 3$
④ $8 : 12 = 2 : 3$
⑤ $3 : 10$

2 ① $1 : 3$ ⑥ $1 : 4$
② $10 : 3$ ⑦ $1 : 3$
③ $3 : 2$ ⑧ $2 : 3$
④ $5 : 2$ ⑨ $3 : 4$
⑤ $6 : 7$ ⑩ $9 : 2$

17 Ratios Ⅵ
P34-35

1 ① (1)

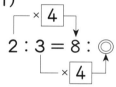

$\circledcirc = 3 \times \boxed{4} = \boxed{12}$

       12 cm

(2) $8 \times \dfrac{3}{2} = \boxed{12}$ 12 cm

② (1)

$$2 : 3 = \circledcirc : 24$$

$\circledcirc = 2 \times \boxed{8} = \boxed{16}$

       16 cm

(2) $24 \times \dfrac{2}{3} = \boxed{16}$ 16 cm

60 ©Kumon Publishing Co., Ltd.

2 ① 48 cm

[Solution 1]

$$5 : 4 = 60 : ◎$$

(×12 →, ×12 ↑)

$$◎ = 4 × 12$$
$$= 48$$

[Solution 2]

$$60 × \frac{4}{5} = 48$$

② 75 cm

[Solution 1]

$$5 : 4 = ◎ : 60$$

(×15 →, ×15 ↑)

$$◎ = 5 × 15$$
$$= 75$$

[Solution 2]

$$60 × \frac{5}{4} = 75$$

3 ① 320 g

[Solution 1]

$$3 : 8 = 120 : ◎$$

(×40 →, ×40 ↑)

$$◎ = 8 × 40$$
$$= 320$$

[Solution 2]

$$120 × \frac{8}{3}$$
$$= 320$$

② 270 g

[Solution 1]

$$3 : 8 = ◎ : 720$$

(×90 →, ×90 ↑)

$$◎ = 3 × 90$$
$$= 270$$

[Solution 2]

$$720 × \frac{3}{8}$$
$$= 270$$

18 **Proportion I** P36-37

1 ① a 2 b 1.5 c 2

② d 3 e 2.5

2 ① (From the left)

8, 12, 16, 20 ✓

② (From the left)

8, 7, 6, 5 ×

③ (From the left)

5, 7.5, 10, 12.5 ✓

④ (From the left)

31, 32, 33, 34 ×

19 **Proportion II** P38-39

1 ① proportional

② (From the left) 6, 6, 6, 6, 6

③ 6

2 Table(From the left) 4,4,4,4

Number sentence

$$Y = 4 × X$$

3 ①

Time X	1	2	3	4	5
Distance Y	50	100	150	200	250
Y ÷ X	50	50	50	50	50

$$Y = 50 × X$$

②

Length X	1	2	3	4	5
Area Y	6	12	18	24	30
Y ÷ X	6	6	6	6	6

$$Y = 6 × X$$

③

Volume X	1	2	3	4	5
Weight Y	0.8	1.6	2.4	3.2	4.0
Y ÷ X	0.8	0.8	0.8	0.8	0.8

$$Y = 0.8 × X$$

20 Proportion Ⅲ　　　P40-41

1 ① $Y = 3 \times X$
　② (1) 30
　　 (2) 45
　　 (3) 180
　　 (4) 600

2 ① $Y = 40 \times X$
　② 100
　③ 8

3 ① $Y = 0.6 \times X$
　② $Y = 3 \times X$
　③ $Y = 65 \times X$

21 Proportion Ⅳ　　　P42-43

1 ① ②

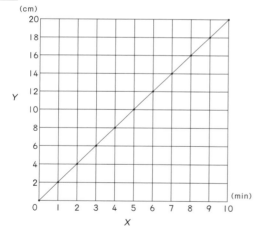

2 ① (From the left)
　　 0.6, 0.9, 1.2, 1.5
　②

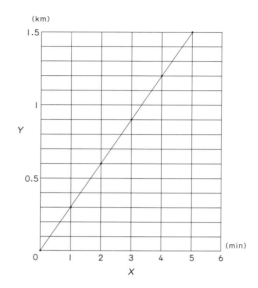

22 Proportion Ⅴ　　　P44-45

1 ① (1) 40 L
　　 (2) 100 L
　② $Y = 20 \times X$
　③ (1) 1 minute
　　 (2) 4 minutes
　④ 50 L

2 ① (1) 10 mi.
　　 (2) 8 mil.
　② Car A $Y = 10 \times X$
　　 Car B $Y = 8 \times X$
　③ Car B

23 Proportion VI P46-47

1 ① Weight of the nails

② (1) $\boxed{50} \div \boxed{25} = \boxed{2}$

$150 \div \boxed{2} = \boxed{75}$

75 nails

(2) $\boxed{150} \div \boxed{50} = \boxed{3}$

$\boxed{3}$

$25 \times \boxed{3} = \boxed{75}$

75 nails

2 ① $60 \div 5 = 12$

$12 \times 24 = 288$

288 g

② $24 \div 5 = 4.8$

$60 \times 4.8 = 288$

288 g

24 Proportion VII P48-49

1 ① Weight of the steel sheet

② (1) $\boxed{100} \div \boxed{125} = \boxed{0.8}$

$\boxed{0.8} \times 225 = \boxed{180}$

180 cm²

(2) $\boxed{225} \div \boxed{125} = \boxed{1.8}$

$\boxed{1.8}$

$100 \times \boxed{1.8} = \boxed{180}$

180 cm²

2 ① $24 \div 15 = 1.6$

$1.6 \times 33 = 52.8$

52.8 cm²

② $33 \div 15 = 2.2$

$24 \times 2.2 = 52.8$

52.8 cm²

25 Review: Proportion P50-51

1 a and d

2 ① $Y = 0.7 \times X$

② $Y = 4 \times X$

③ $Y = 3.5 \times X$

3 ① $Y = 8 \times X$

② 80 g

③ 200 g

④ 30 cm³

⑤ 75 cm³

26 Review P52-53

1 ① $150 \div 24 = 6.25$

6.25 m/sec

② $6.25 \times 60 = 375$

375 m/min

2 ① 2 : 4

② 4 : 6

③ 8 : 6

④ 20 : 12

3 ①

Length	X	0	1	2	3	4	5
Cost	Y	0	1.40	2.80	4.20	5.60	7.00

② 1.40

③ $Y = 1.40 \times X$

④ $1.40 \times 7 = 9.80$

$\$ 9.80$

27 Review

P54-55

1 ① 9

② 9

③ 25

④ 9

⑤ 36

⑥ 4

⑦ 4

⑧ 4

⑨ 3

⑩ 3

2 $800 \div 480$

$= \dfrac{800}{480} = \dfrac{5}{3} = 1\dfrac{2}{3}$ h

$60 \times \dfrac{2}{3} = 40$ min

1 hour 40 minutes

3 ①

Length	X	0	1	2	3	4	5
Cost	Y	0	2.00	4.00	6.00	8.00	10.00

②

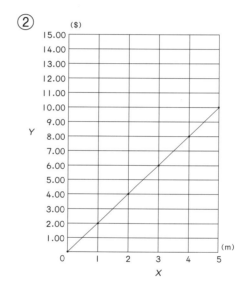

③ $Y = 2.00 \times X$